Differentiated Worksheets (K-6) SCIENCE, VOL 1

Created By: Tavio M. Soto

Cover Art:

Tavio M. Soto

Contact:

For information about this book or to reach the author, email – TavioSotoOfficial@gmail.com. Also, visit my website at TavioSoto.com.

--- TABLE OF CONTENTS ---

MEET THE
Teacher

About Me

My name is Tavio Soto and I'm a passionate and dedicated Special Education teacher, retired military firefighter, and an Amazon Author.

Education

- EC-12 Special Education & EC-6 Generalist Cert'd
- Masters Degree in Executive Leadership
- Bachelors Degree in Religion
- Louis F. Garland Fire Academy Master Instructor

Amazon Books

Pupster
Pupster II - Saving Sarah
Workbooks

Fun Fact

I post free content on Teacher's Pay Teacher's to help teachers thrive!

 605.391.3080   TavioSotoOfficial@gmail.com

The WHAT

This workbook was created out of a need that I identified over several years of working in the classroom. The content covers areas where many students struggle.

The WHY

This workbook also addresses a need for teachers to have differentiated assignments when working on a single topic. It's a work smarter, not harder concept to make your life a little easier.

The HOW

The rigor gradually increases as you go deeper into the subject matter, giving you tools to use for every level of student in your classroom.

SCIENCE WORKSHEETS

air

A mixture of gases (especially oxygen) that is required for breathing.

1. **Trace the letters.**

2. **Circle the letters - a / i / r.**

d f t a o v l w p i z t n r o

3. **Write the word, <u>air</u>, six times. (lowercase)**

_____

_____

_____

animal

A living organism that eats, breaths, and moves.

1. **Trace the letters.**

animal

2. **Circle the letters - a / n / i / m / a / l.**

d f t a o n i w p m a t c l

3. **Write the word, <u>animal</u>, six times. (lowercase)**

_____

_____

_____

Name: _____ **Date:** _____

acid

A corrosive substance.

1. **Trace the letters.**

2. **Circle the letters - a / c / i / d.**

x f t a o n c w i m a t n d

3. **Write the word, <u>acid</u>, six times.** (lowercase)

_____

_____

_____

Name: _____ **Date:** _____

astronaut

A person who explores the regions of our outer atmosphere.

1. **Trace the letters.**

astronaut

2. **Circle the letters - a / s / t / r / o / n / a / u / t.**

a u r s l t e c r o c n a u t

3. **Write the word, astronaut, six times. (lowercase)**

_____

_____

_____

astronomy

The study of everything in the universe beyond earth's atmosphere.

1. **Trace the letters.**

astronomy

2. **Circle the letters - a / s / t / r / o / n / o / m / y.**

a u r s l t e c r o n i o m y

3. **Write the word, <u>astronomy</u>, six times. (lowercase)**

_____

_____

_____

aquarium

Public access ocean life reserve.

1. **Trace the letters.**

aquarium

2. **Circle the letters - a / q / u / a / r / i / u / m.**

t a r q l u e a t r c i u e m

3. **Write the word, <u>aquarium</u>, six times. (lowercase)**

_____

_____

_____

base

A substance that can neutralize an acid.

1. **Trace the letters.**

base

2. **Circle the letters – b / a / s / e.**

b l t a s r e m p i f t n c h

3. **Write the word, <u>base</u>, six times. (lowercase)**

_____

_____

_____

branch

The part of a plant that extends from the stem or trunk.

1. **Trace the letters.**

branch

2. **Circle the letters – b / r / a / n / c / h.**

s l t e b r a m p i f t n c h

3. **Write the word, <u>branch</u>, six times. (lowercase)**

_____

_____

_____

burn

Chemical reaction that creates heat energy and light.

1. **Trace the letters.**

burn

2. **Circle the letters - b / u / r / n.**

s f t a b v l u p r o t n c w

3. **Write the word, <u>burn</u>, six times. (lowercase)**

_____

_____

_____

carnivore

An animal that eats the flesh of other animals.

1. **Trace the letters.**

carnivore

2. **Circle the letters - c / a / r / n / i / v / o / r / e.**

t c a s l r z n m i v o l r e

3. **Write the word, <u>carnivore</u>, six times. (lowercase)**

_____

_____

_____

climate

The weather conditions prevailing in an area in general over a longer period of time.

1. **Trace the letters.**

climate

2. **Circle the letters - c / l / i / m / a / t / e.**

t c r s l a i c m y a t l e b

3. **Write the word, <u>climate</u>, six times. (lowercase)**

_____

_____

_____

cloud

A mixture of gas particles that you can see.

1. **Trace the letters.**

2. **Circle the letters - c / l / o/ u / d.**

d c t a l v o w p u a t d z

3. **Write the word, <u>cloud</u>, six times. (lowercase)**

_____

_____

_____

Name: _____ **Date:** _____

cold

Having little or no warmth.

1. **Trace the letters.**

2. **Circle the letters - c / o / l / d.**

s c t a n o l m p i d t n x

3. **Write the word, <u>cold</u>, six times. (lowercase)**

_____

_____

_____

Name: _____ **Date:** _____

color

Color is what you see when light reflects off an object.

1. **Trace the letters.**

2. **Circle the letters - c / o / l / o / r.**

t u c s q a e o t y l i o e r

3. **Write the word, <u>color</u>, six times. (lowercase)**

_____

_____

_____

curved

A direction with any directional degree of change.

1. **Trace the letters.**

curved

2. **Circle the letters - c / u / r / v / e / d.**

c u r s l a v e t y k i l d b

3. **Write the word, <u>curved</u>, six times. (lowercase)**

_____

_____

_____

Name: _____ **Date:** _____

data

Information that is gathered and stored.

1. **Trace the letters.**

2. **Circle the letters - d / a / t / a.**

d u r s l a e c t y c a l e b

3. **Write the word, <u>data</u>, six times. (lowercase)**

_____

_____

_____

Name: _____ **Date:** _____

day

The duration of time that the sun lights up an area.

1. **Trace the letters.**

2. **Circle the letters - d / a / y.**

d f t a o v l w p i y t n c

3. **Write the word, <u>day</u>, six times. (lowercase)**

_____

_____

_____

Name: _____ **Date:** _____

desert

A hot dry climate that has little water.

1. **Trace the letters.**

2. **Circle the letters - d / e / s / e / r / t.**

d u e s l a e c r y c i t p b

3. **Write the word, <u>desert</u>, six times. (lowercase)**

_____

_____

_____

Name: _____ **Date:** _____

dormant

The process of plants and some animals having normal functions slowed down during a season.

1. **Trace the letters.**

dormant

2. **Circle the letters - d / o / r / m / a / n / t.**

d u o s l r e c m y a n t e

3. **Write the word, <u>dormant</u>, six times. (lowercase)**

_____

_____

_____

earth

The planet on which we live.

1. **Trace the letters.**

2. **Circle the letters - e / a / r / t / h.**

m u r s l x e c a y r i t h b

3. **Write the word, <u>earth</u>, six times. (lowercase)**

_____

_____

_____

Name: _____ **Date:** _____

eat

The process of chewing and swallowing nutritious substances.

1. **Trace the letters.**

2. **Circle the letters - e / a / t.**

m u r s l a e c a y c i t h b

3. **Write the word, <u>eat</u>, six times. (lowercase)**

_____

_____

_____

electricity

The flow of electrical power or charge.

1. **Trace the letters.**

electricity

2. **Circle the letters - e / l / e / c / t / r / i / c / i / t / y.**

e u l s e c t r a i c i t h y

3. **Write the word, <u>electricity</u>, six times. (lowercase)**

_____

_____

_____

fall

The season is characterized by the changing colors of leaves and temperatures dropping.

1. **Trace the letters.**

2. **Circle the letters - f / a / l / l.**

s f t a n v l m p l r t n e

3. **Write the word, <u>fall</u>, six times. (lowercase)**

_____

_____

_____

Name: _____ **Date:** _____

fire

The heat and light that is the direct result of heating a combustible material.

1. **Trace the letters.**

2. **Circle the letters - f / i / r / e.**

s f t a n v l m p i r t n e

3. **Write the word, _fire_, six times. (lowercase)**

_____

_____

_____

food

Any nutritious substance that people or animals eat or drink or that plants absorb.

1. **Trace the letters.**

2. **Circle the letters - f / o / o / d.**

t u r s f a o c o y c i l e d

3. **Write the word, <u>food</u>, six times. (lowercase)**

_____

_____

_____

Name: _____ **Date:** _____

forest

An area of land that has lots of trees and green vegetation.

1. **Trace the letters.**

2. **Circle the letters - f / o / r / e / s / t.**

t u f o l a r c e y s i t e b

3. **Write the word, <u>forest</u>, six times. (lowercase)**

_____

_____

_____

Name: _____ **Date:** _____

fossil

A mineralized form of a plant or animal, often found in rocks.

1. **Trace the letters.**

2. **Circle the letters - f / o / s / s / i / l.**

t u r p f a o c s s c i l e b

3. **Write the word, <u>fossil</u>, six times. (lowercase)**

_____

_____

_____

fruit

The part of a plant that grows for the purpose of being eaten by animals.

1. **Trace the letters.**

2. **Circle the letters - f / r / u / i / t.**

s l f e n r a m u i p t n c

3. **Write the word, <u>fruit</u>, six times. (lowercase)**

_____

_____

_____

herbivore

Animals that only eat vegetation.

1. **Trace the letters.**

herbivore

2. **Circle the letters - h / e/ r / b / i / v / o / r / e.**

h i e r l k b c i y v s o r e

3. **Write the word, <u>herbivore</u>, six times. (lowercase)**

_____

_____

_____

hibernate

The action of an animal or plant spending the winter in a dormant state.

1. **Trace the letters.**

hibernate

2. **Circle the letters - h / i / b / e / r / n / a / t / e.**

h i y b l e r c n y a s t i e

3. **Write the word, <u>hibernate</u>, six times. (lowercase)**

_____

_____

_____

Name: _____ **Date:** _____

hypothesis

Making an educated guess about the expected results of an experiment.

1. **Trace the letters.**

hypothesis

2. **Circle the letters - h / y / p / o / t / h / e / s / i / s.**

huyplotchyeslis

3. **Write the word, <u>hypothesis</u>, six times. (lowercase)**

_____

_____

_____

Name: _____ **Date:** _____

lake

It's a body of fresh water that is contained by large land areas. Smaller than the ocean, bigger than a pond.

1. **Trace the letters.**

lake

2. **Circle the letters - l / a / k / e.**

t u r s l a k c t y c i j e b

3. **Write the word, <u>lake</u>, six times. (lowercase)**

_____

_____

_____

laws

The description of naturally occurring phenomena in the world, that is always constant.

1. **Trace the letters.**

2. **Circle the letters - l / a / w / s.**

t u r x l a e c w y c i s e

3. **Write the word, <u>laws</u>, six times. (lowercase)**

_____

_____

_____

Name: _____ **Date:** _____

leaf

An extension of a plant that produces oxygen and collects carbon dioxide.

1. **Trace the letters.**

2. **Circle the letters - l / e / a / f.**

s l t e n v a m p i f t n c

3. **Write the word, _leaf_, six times. (lowercase)**

_____

_____

_____

lightning

An atmospheric reaction caused by the collision of positive and negative charged particles.

1. **Trace the letters.**

lightning

2. **Circle the letters - l / i / g / h / t / n / i / n / g.**

s l i e g h a m p t f n i n g

3. **Write the word, _lightning_, six times. (lowercase)**

_____

_____

_____

Name: _____ **Date:** _____

magnet

A piece of metal with a strong attraction to another metal object.

1. **Trace the letters.**

magnet

2. **Circle the letters - m / a / g / n / e / t.**

m a g n l v e c t y c i l f b

3. **Write the word, <u>magnet</u>, six times. (lowercase)**

_____

_____

_____

magnetic

Able to be attracted by a magnet.

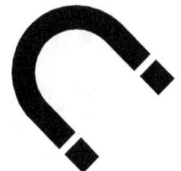

1. **Trace the letters.**

magnetic

2. **Circle the letters - m / a / g / n / e / t / i / c.**

m u a s g n e r t y s i l c b

3. **Write the word, <u>magnetic</u>, six times. (lowercase)**

_____

_____

_____

mass

The amount of matter or substance that makes up an object.

1. **Trace the letters.**

2. **Circle the letters - m / a / s / s.**

t u r m l a e c t y s i l s b

3. **Write the word, <u>mass</u>, six times. (lowercase)**

_____

_____

_____

Name: _____ **Date:** _____

matter

Anything that takes up space and can be weighed.

1. **Trace the letters.**

2. **Circle the letters - m / a / t / t / e / r.**

m u r s l a e c t y t i l e r

3. **Write the word, <u>matter</u>, six times. (lowercase)**

_____

_____

_____

meteorite

Particles of matter that fall through the atmosphere.

1. **Trace the letters.**

meteorite

2. **Circle the letters - m / e / t / e / o / r / i / t / e.**

m u e t l a e o r y c i t e b

3. **Write the word, <u>meteorite</u>, six times. (lowercase)**

_____

_____

_____

meteor shower

Several particles of matter that fall through the atmosphere during a close period of time.

1. **Trace the letters.**

meteor shower

2. **Circle the letters - m / e / t / e / o / r / s / h / o / w / e / r.**

meteoreshowler

3. **Write the word, <u>meteor shower</u>, (3) times. (lowercase)**

_____

_____

_____

microscope

An instrument that is used to magnify small objects.

1. **Trace the letters.**

microscope

2. **Circle the letters - m / i / c / r / o / s / c / o / p / e.**

microvlmpscope

3. **Write the word, <u>microscope</u>, six times. (lowercase)**

_____

_____

_____

Name: _____ **Date:** _____

migration

The seasonal movement of animals from one region to another.

1. **Trace the letters.**

migration

2. **Circle the letters - m / i / g / r / a / t / i / o / n.**

m i c r o g l r p a t i o n

3. **Write the word, <u>migration</u>, six times. (lowercase)**

_____

_____

_____

moon

Earth's natural satellite. Seen at night. Goes through different phases.

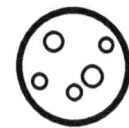

1. **Trace the letters.**

2. **Circle the letters - m / o / o / n.**

d m a o v l w o i a t n c y

3. **Write the word, <u>moon</u>, six times. (lowercase)**

_____

_____

_____

motion

Any type of movement.

1. **Trace the letters.**

2. **Circle the letters - m / o / t / i / o / n.**

m u r s l a o c t y c i o e n

3. **Write the word, motion, six times. (lowercase)**

_____

_____

_____

natural resource

Any item used by man that is produced in nature.

1. **Trace the letters.**

natural resource

2. **Circle - n / a / t / u / r / a / l / r / e / s / o / u / r / c / e.**

naturalresource

3. **Write the word, natural resource, (3) times. (lowercase)**

_____

_____

_____

Name: _____ **Date:** _____

night

The duration of time where the sunlight is not present in an area.

1. **Trace the letters.**

2. **Circle the letters - n / i / g / h / t.**

d f n a o v l w p i g h c t

3. **Write the word, <u>air</u>, six times. (lowercase)**

_____

_____

_____

Name: _____ **Date:** _____

observe

Making a purposeful action to observe something for study.

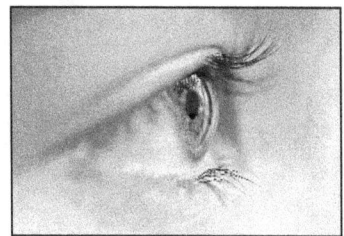

1. **Trace the letters.**

observe

2. **Circle the letters - o / b / s / e / r / v / e.**

o b r s l a e c t r c v l e b

3. **Write the word, <u>observe</u>, six times. (lowercase)**

_____

_____

_____

Name: _____ **Date:** _____

ocean

The largest body of water on earth. It is salty.

1. **Trace the letters.**

2. **Circle the letters - o / c / e / a / n.**

t u r o l t n c t y e i l a n

3. **Write the word, <u>ocean</u>, six times. (lowercase)**

_____

_____

_____

omnivore

An animal that eats both meat and vegetation.

1. **Trace the letters.**

omnivore

2. **Circle the letters - o / m / n / i / v / o / r / e.**

turolmentiviore

3. **Write the word, <u>omnivore</u>, six times. (lowercase)**

_____

_____

_____

planet

Large spherical objects outside of Earth's atmosphere.

1. **Trace the letters.**

2. **Circle the letters - p / l / a / n / e / t.**

m u r p l a f n e y c t l k

3. **Write the word, <u>planet</u>, six times. (lowercase)**

_____

_____

_____

planting

The process of growing a plant from a seed or seedling by placing it in soil.

1. **Trace the letters.**

planting

2. **Circle the letters - p / l / a / n / t / i / n / g.**

t y p s l r a o n c t i n g

3. **Write the word, <u>planting</u>, six times. (lowercase)**

_____

_____

_____

pollution

The introduction of harmful substances into a clean environment.

1. **Trace the letters.**

pollution

2. **Circle the letters - p / o / l / l / u / t / k / i / o / n.**

p o w s l a e l u t d i o n

3. **Write the word, <u>pollution</u>, six times. (lowercase)**

_____

_____

_____

rain

The collection of water particles that fall from clouds.

1. **Trace the letters.**

2. **Circle the letters - r /a / i / n.**

d f r a i n l w p p w t l c

3. **Write the word, <u>rain</u>, six times. (lowercase)**

_____

_____

_____

Name: _____ **Date:** _____

research

Gathering information about a subject or process.

1. **Trace the letters.**

research

2. **Circle the letters - r / e / s / e / a / r / c / h.**

t u r v l e s o t e a i r c h

3. **Write the word, <u>research</u>, six times. (lowercase)**

_____

_____

_____

Name: _____ **Date:** _____

rocket

A cylindrical-shaped vehicle used to travel to high altitudes.

1. **Trace the letters.**

rocket

2. **Circle the letters - r / o / c / k / e / t.**

x u r s o a e c b y k i l e t

3. **Write the word, <u>rocket</u>, six times. (lowercase)**

_____

_____

_____

roots

The part of the plant that grows into the ground that absorbs nutrients and water.

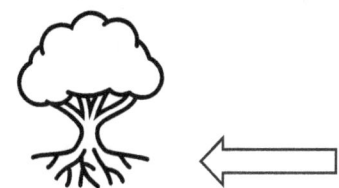

1. **Trace the letters.**

2. **Circle the letters - r / o / o / t / s.**

g f a n v l r p o o t n c s

3. **Write the word, <u>roots</u>, six times. (lowercase)**

_____

_____

_____

rotate

Motion that is typically viewed as a three-hundred-and-sixty-degree transition. Changing positions.

1. **Trace the letters.**

2. **Circle the letters - r / o / t / a / t / e.**

c u r s l o s c t a c i t e b

3. **Write the word, <u>rotate</u>, six times. (lowercase)**

_____

_____

_____

rough

A surface that is not smooth.

1. **Trace the letters.**

2. **Circle the letters - r / o / u / g / h.**

t u r s l o e c u y g i h e b

3. **Write the word, <u>rough</u>, six times. (lowercase)**

_____

_____

_____

Name: _____ **Date:** _____

ruler

An instrument used to measure things utilizing inches or centimeters.

1. **Trace the letters.**

ruler

2. **Circle the letters - r / u / l / e / r.**

t v r s l o e c u y l i h e r

3. **Write the word, <u>ruler</u>, six times. (lowercase)**

_____

_____

_____

satellite

An aerial device that transmits and receives information to and from the surface of the earth.

1. **Trace the letters.**

2. **Circle the letters - s / a / t / e / l / l / i / t / e.**

s u a s t e l c l y c i t e b

3. **Write the word, <u>satellite</u>, six times. (lowercase)**

_____

_____

_____

Name: _____ **Date:** _____

scale

An instrument that is used to measure weight.

1. **Trace the letters.**

2. **Circle the letters - s / c / a / l / e.**

s f t c n v a m p l o t e z

3. **Write the word, <u>scale</u>, six times. (lowercase)**

_____

_____

_____

scientist

An expert in science.

1. **Trace the letters.**

scientist

2. **Circle the letters - s / c / i / e / n / t / i / s / t.**

t u r s c i e n t y c i s e t

3. **Write the word, <u>scientist</u>, six times. (lowercase)**

_____

_____

_____

Name: _____ **Date:** _____

seasons

Four distinctly different time periods during the year with different types of weather.

1. **Trace the letters.**

seasons

2. **Circle the letters - s / e / a / s / o / n / s.**

t u r s l e a c s y o n l s b

3. **Write the word, <u>seasons</u>, six times. (lowercase)**

_____

_____

_____

Name: _____ **Date:** _____

seeds

The part of the plant that is very small and carries the genetic information to grow a new plant.

1. **Trace the letters.**

2. **Circle the letters - s / e / e / d / s.**

t y p s l r e o n e d i s g

3. **Write the word, <u>seeds</u>, six times. (lowercase)**

_____

_____

_____

Name: _____ **Date:** _____

sky

The upper atmosphere of the earth.

1. **Trace the letters.**

2. **Circle the letters - s / k / y.**

s f t a k v l w p i y t n c o

3. **Write the word, <u>sky</u>, six times. (lowercase)**

_____

_____

_____

Name: _____ **Date:** _____

smooth

A surface that is absent of a rough feeling.

1. **Trace the letters.**

2. **Circle the letters - s / m / o / o / t / h.**

t u r s l a m c o o c t h e

3. **Write the word, <u>smooth</u>, six times. (lowercase)**

_____

_____

_____

snow

The collection of frozen water particles that falls from clouds.

1. **Trace the letters.**

2. **Circle the letters - s / n / o / w.**

s f t a n l m p i o t n c w

3. **Write the word, <u>snow</u>, six times. (lowercase)**

_____

_____

_____

soil

A mixture of materials that are found on the ground that plants use for rooting and food.

1. **Trace the letters.**

2. **Circle the letters - s / o / i / l.**

t y f s l r e o m c t i p l

3. **Write the word, <u>soil</u>, six times. (lowercase)**

_____

_____

_____

sound

Vibrations created by environmental motion that are heard by a person's ear.

1. **Trace the letters.**

sound

2. **Circle the letters - s / o / u / n / d.**

t u r s l o e u t y n d l e b

3. **Write the word, <u>sound</u>, six times. (lowercase)**

_____

_____

_____

space shuttle

A vehicle that is used to explore space.

1. **Trace the letters.**

space shuttle

2. **Circle the letters - s / p / a / c / e / s / h / u / t / t / l / e.**

s p r s l a c e s h u t t l e

3. **Write the word, <u>space shuttle</u>, (3) times. (lowercase)**

_____

_____

_____

spacesuit

It's like a wet suit that provides protection and oxygen for astronauts.

1. **Trace the letters.**

spacesuit

2. **Circle the letters - s / p / a / c / e / s / u / i / t.**

s u p a l c e s z y u i l t b

3. **Write the word, <u>spacesuit</u>, six times. (lowercase)**

_____

_____

_____

spring

The season is characterized by new growth, rain, and flowers. The season after winter but before summer.

1. **Trace the letters.**

spring

2. **Circle the letters - s / p / r / i / n / g.**

d u m s p a e r i y n h g

3. **Write the word, spring, six times. (lowercase)**

_____

_____

_____

straight

A direction characterized by an absence of a curve.

1. **Trace the letters.**

2. **Circle the letters - s / t / r / a / i / g / h / t.**

d u k s t r a c i y g h l t b

3. **Write the word, <u>straight</u>, six times. (lowercase)**

_____

_____

_____

summer

The hottest season of the year.

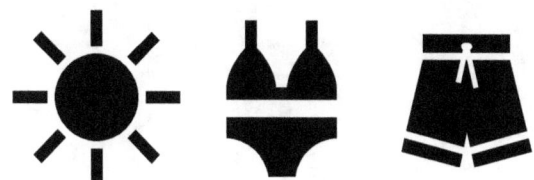

1. **Trace the letters.**

summer

2. **Circle the letters - s / u / m / m / e / r.**

d f s a o u l m m i e t r c

3. **Write the word, <u>summer</u>, six times. (lowercase)**

_____

_____

_____

sun

The brightest round object in the sky that gives the earth light and heat.

1. **Trace the letters.**

sun

2. **Circle the letters - s / u / n.**

d f s a o v l w u i a t n c

3. **Write the word, <u>sun</u>, six times. (lowercase)**

_____

_____

_____

technology

Any device created from natural resources that is used as a tool.

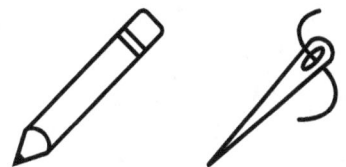

1. **Trace the letters.**

technology

2. **Circle the letters - t / e / c / h / n / o / l / o / g / y.**

t y f e c h n m s o l o g y

3. **Write the word, <u>technology</u>, six times. (lowercase)**

_____

_____

_____

165

Name: _____ **Date:** _____

telescope

An instrument used to view objects far away, like the stars.

1. **Trace the letters.**

telescope

2. **Circle the letters - t / e / l / e / s / c / o / p / e.**

t y f e l r e m s c t o p e

3. **Write the word, <u>telescope</u>, six times. (lowercase)**

_____

_____

_____

thermometer

A tool that is used to measure temperature.

1. **Trace the letters.**

thermometer

2. **Circle the letters - t / h / e / r / m / o / m / e / t / e / r.**

f t h e r m o p f m e t e r

3. **Write the word, _thermometer_, six times. (lowercase)**

_____

_____

_____

thunder

The booming sound that often occurs after a strike of lightning.

1. **Trace the letters.**

thunder

2. **Circle the letters - t / h / u / n / d / e / r.**

s f t h c r u o n f d e i b r

3. **Write the word, <u>thunder</u>, six times. (lowercase)**

_____

_____

_____

Name: _____ **Date:** _____

time

Measurement of how long it takes for things to happen.

1. **Trace the letters.**

2. **Circle the letters - t / i / m / e.**

s f t h i r m o p f b e h x r

3. **Write the word, <u>time</u>, six times. (lowercase)**

_____

_____

_____

tornado

A destructive force of nature that has the appearance of a funnel-shaped cloud.

1. **Trace the letters.**

tornado

2. **Circle the letters - t / o / r / n / a / d / o.**

t y o r l n e a t d o i r g

3. **Write the word, tornado, six times. (lowercase)**

_____

_____

_____

trash

The part of an item or material that is no longer valuable to a person and is considered a waste.

1. **Trace the letters.**

trash

2. **Circle the letters - t / r / a / s / h.**

t y w r l a e s b e h i r g

3. **Write the word, <u>trash</u>, six times. (lowercase)**

_____

_____

_____

Name: _____ **Date:** _____

tree

A plant that has a woody main stem or trunk that usually grows to a considerable height.

1. **Trace the letters.**

2. **Circle the letters - t / r / e / e.**

s f t a n v r m q e e z n c

3. **Write the word, <u>tree</u>, six times. (lowercase)**

_____

_____

_____

volume

A measurement referring to the loudness of sound waves.

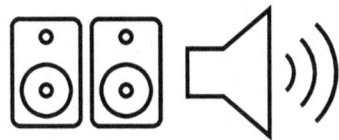

1. **Trace the letters.**

volume

2. **Circle the letters - v / o / l / u / m / e.**

v u o s l a b c t u c i m e

3. **Write the word, <u>volume</u>, six times. (lowercase)**

_____

_____

_____

Name: _____ **Date:** _____

water

A transparent and odorless liquid that all living things need to exist.

1. **Trace the letters.**

2. **Circle the letters - w / a / t / e / r.**

c y w s l a u o t e d i r g

3. **Write the word, <u>water</u>, six times. (lowercase)**

_____

_____

_____

weight

The measure of the force of gravity on an object.

1. **Trace the letters.**

weight

2. **Circle the letters - w / e / i / g / h / t.**

c y w s l a e i g p h t r z

3. **Write the word, <u>weight</u>, six times. (lowercase)**

_____

_____

_____

winter

The coldest season of the year.

1. **Trace the letters.**

winter

2. **Circle the letters - w / i / n / t / e / r.**

l y w s i a n t g p h e r z

3. **Write the word, <u>winter</u>, six times. (lowercase)**

_____

_____

_____

yardstick

An instrument used to measure up to three feet in length.

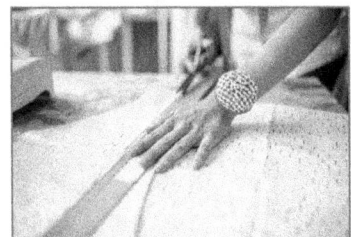

1. **Trace the letters.**

yardstick

2. **Circle the letters - y / a / r / d / s / t / i / c / k.**

y u a n r d e o s t i o c k

3. **Write the word, <u>yardstick</u>, six times. (lowercase)**

_____

_____

_____

ZOO

Public access animal reserve.

1. **Trace the letters.**

zoo

2. **Circle the letters - z / o / o.**

t u r s z a e o t y c o l e b

3. **Write the word, <u>zoo</u>, six times. (lowercase)**

_____

_____

_____

zoology

The scientific study of animals.

1. **Trace the letters.**

zoology

2. **Circle the letters - z / o / o / l / o / g / y.**

t z r o o a e l t o c g x y

3. **Write the word, <u>zoology</u>, six times. (lowercase)**

_____

_____

_____

Final Thoughts

I hope this resource helps you reach your students in an impactful and encouraging way. May your students grow stronger and healthier daily because you are in their lives.

I teach special education, I love it! To help supplement my income, I'm also a children's book author. I've written a two-book series (Pupster & Pupster II) that is helping to bring back the joy of reading to kids. These books also teach invaluable life lessons. Please take the time to check them out. Thank you for considering them for your school, family and friends.

Much Love,

Tavio M. Soto

Written for youth between the ages of 10 and 16. Pupster teaches invaluable values while helping students stay engaged with the critical skill of reading.

Get Your Copy of Pupster on Amazon!

Pupster II, Saving Sarah is the continuation of the Pupster saga. It is written to teach kids about the dangers of internet predators and how to survive a kidnapping situation.

Available on Amazon!

Pupster is also available in Spanish.

Look for it on Amazon!

PUPSTER

Versión en Español

Tavio M. Soto

TAVIO M. SOTO

SCAN QR CODE TO GO STRAIGHT TO AMAZON!